Alchemy For]

Discover the Magic and Mystery of the Ancient Alchemy Craft and Use It to Your Advantage

Alchemist Tony

Your Free Gift

As a way of thanking you for the purchase, I'd like to offer you a complimentary gift:

- **5 Pillar Life Transformation Checklist:** This short book is about life transformation, presented in bit size pieces for easy implementation. I believe that without such a checklist, you are likely to have a hard time implementing anything in this book and any other thing you set out to do religiously and sticking to it for the long haul. It doesn't matter whether your goals relate to weight loss, relationships, personal finance, investing, personal development, improving communication in your family, your overall health, finances, improving your sex life, resolving issues in your relationship, fighting PMS successfully, investing, running a successful business, traveling etc. With a checklist like this one, you can bet that anything you do will seem a lot easier to implement until the end. Therefore, even if you don't continue reading this book, at least read the one thing that will help you in every other aspect of your life. Grab your copy now by clicking/tapping here or simply enter http://bit.ly/2fantonfreebie into your browser. Your life will never be the same again (if you implement what's in this book), I promise.

PS: I'd like your feedback. If you are happy with this book, please leave a review on Amazon.

Introduction

Thinking about transformation? How many times has this been the baseline of your goals? You know that to achieve that (a goal), you must stop doing this (some destructive habits); you must transform. It happens in many levels; there is a transformation of the mind (beliefs, thoughts and perceptions) to transformation of a livelihood.

It is not something that will happen overnight when you read a 'life transforming' book. It is a process which starts up in our minds, melting onto our soul and spirit and out flowing in our day to day life; habits and actions. It is unfortunate that most of us never get to transform because we do not understand these facts.

When you think about changing your life for better, think about Alchemy. It was an ancient science where a few scientists went to pains to make gold out of lead. The believed that the former was not just a worthless metal; the only reason why it was not as appealing as the finest metal, was simply because it had not developed to that stage. They strongly believed that if a few changes were made to it, then it would become gold. They had quite the growth mindset, right?

Today, we cannot try to change metals, but we can borrow very useful knowledge from their practice to improve our lives today. They never believed any metal was worthless in the same way that none of us is. No matter how 'dull' you are, it does not mean that you are to be written off. It is possible to be polished to become worthy and of value, just like gold.

The modern alchemy is not about metals. It is about the transformative processes of the human species. In his book the "Alchemist", Paul Coelho says it best *"real alchemy in the world is the ability to manifest personal change – an ability that everyone possesses"*.

You and anyone else who wants to can change to be the best version of themselves. Ancient alchemists may never have managed to change lead to gold but they did leave us some wisdom in the 7 stages of alchemy, which we can use to study about transformation for 'dramatic self-improvement' as Paul calls it. This book introduces you to alchemy. The book will guide you through the process and teach you how to become a modern day alchemist so you can improve the quality of your life, live purposefully and gain success and complete happiness.

Alchemy For Beginners

© Copyright 2019 Alchemist Tony - All rights reserved.

Table of Contents

Your Free Gift 2

Introduction 3

Chapter 1: An Introduction To Alchemy ___ 8

Physical Alchemy 8

Spiritual Alchemy 9

How Real Is Alchemy? 10

Chapter 2: Is Spiritual Alchemy The 'Healer' Of Our World? 12

The Scam That Is Science, The Rational Mind And Technology 12

Chapter 3: Application Of Alchemy _____ 16

The Seven Stages Of Spiritual Alchemy _____ 17

Chapter 4: Effecting Transformation With The 7 Stages Of Alchemy In Modern Life _ 36

Breaking Down The Present Self _____ 36

Creating A New Self _____ 47

Chapter 5: Keep Your Alchemist Practice Alive And Maintain The New Sense Of Self 53

Conclusion _____ **61**

Do You Like My Book & Approach To Publishing? _____ **62**

   1: First, I'd Love It If You Leave a Review of This Book on Amazon._____ 62

   2: Grab Some Freebies On Your Way Out; Giving Is Receiving, Right? _____ 62

PSS: Let Me Also Help You Save Some Money! _____ **63**

Chapter 1: An Introduction To Alchemy

It sounds like some sophisticated kind of science, right? I thought so too until I dug a little deeper. I found that it is some kind of ancient science, which defined many people's lives and whose disciplines we can borrow from to help us create a better and higher life experience.

Physical Alchemy

To begin with, Alchemy is an ancient practice that was shrouded in mystery and secrecy. The main aim of those who practiced it was to turn lead (the most basic metal) to gold (the most precious metal).

It is understandable why there were many secrets and mysteries among the practitioners – I mean, who would want anyone else to find out how they could make such a precious metal.

Thankfully, the ancient world is not like the world of today where people do thing mostly for material gain. Ancient alchemists did not seek to change base metals like lead to gold purely out of greed to get rich. There was something deeper about it; they believed they were developing an 'immature' metal to reach its highest development in nature.

The thing is; they believed that metal is not dead matter. According to ancient alchemists (and their followers), metals are alive and they grow inside the earth. What's more, they have stages of growth such as the human stages of childhood, teenage and adulthood.

So, when they found a base metal such as lead, it was not thought of as unworthy or to be of low quality. It was viewed as physically and spiritually immature version of the high quality metals such as gold.

These alchemists believed that no metal existed as a unique substance; that all were simply different versions of the same thing made different by the fact that they were in different stages of development (or refinement). So, this is why they took it upon themselves to develop or refine a metal to become the enviable and precious metal.

Gold was believed to be a metal that has reached its highest level of development, and they sought to bring those base metals to this development stage. Did they succeed? There are no records of any of them succeeding but who knows. It is said that most of the alchemist literature was destroyed.

We only just talked about the physical alchemy. There is another part to it;

Spiritual Alchemy

The spiritual alchemists were looking to find the 'holy grail'; this was their 'gold'. The Holy Grail refers to uniting of the mind (physical body) with the soul. An individual was considered to be 'golden' if their physical being was in sync with their spiritual being (soul) and had defeated the power of evil.

They were endowed with spiritual beauty. An individual who was disconnected from their soul and was engulfed by evil and stuck with their wayward ways was considered a base metal, lead.

How Real Is Alchemy?

The physical part, the one for changing lead to gold is not real; it is impossible to achieve and alchemists who tried it failed miserably. They based their theories and experiments on a misguided Aristotelian assumption that everything in the world is composed on 4 basic elements namely fire, air, water and earth along with other essential substances namely mercury, sulphur and salt. The percentages of these elements could be adjusted to transform a metal from a base metal to the fine quality metals.

Today, we know that the universe is made up of just elements and atoms. They cannot be adjusted to turn those already in existence to anything else. Lead cannot be and has never been able to turn to gold.

They were wrong but not completely out of it. Here is the thing; today it is possible to create gold from other elements using the modern physics equipment like the particle accelerator. However, the resulting amounts are said to be sub-microscopic, not to mention that the costs exceed the quality of the gold. This is simply to say that no one would want to do this – it's not worth it.

The spiritual part however, is alive today. It is the basis of the beliefs of Enlightment and illumination by the people of the east (in Buddhism). They have spread throughout the world and even adapted by the western cultures that now practice meditation, yoga, mindful breathing and so on all in an effort to connect the soul and the physical body in search of Enlightment and clarity.

So, Physical alchemy may have died, but the spiritual teaching of it never left us. Actually, alchemy can be said to be a type of transformation; going through a process of identification, separation and molding to be changed or to be developed from one form and be made to another higher form. Therefore, we can conclude that the true treasure of alchemy is achieving the golden state; the state of Enlightment and illumination.

I know you might be thinking; how can this impact or influence your day to day life? What can you borrow from the lessons of alchemy and how can you apply them? Let's get to that in the next chapter.

Chapter 2: Is Spiritual Alchemy The 'Healer' Of Our World?

Alchemy is more than a failed science and a mysterious misguided get rich quick scheme of turning a worthless and dangerous metal into something as precious as gold.

Changing a low quality metal to gold was not just that. It is symbolism of spiritual upgrade, to live life in the highest capacity possible as a human being after going through a process which drives us to evolve into refined and elegant beings.

Alchemy is an ancient tradition that looked to transform ordinary people into extraordinary human beings i.e. a 'normal' life to a spiritually rich life directed towards a divine purpose and lived with passion.

The Scam That Is Science, The Rational Mind And Technology

In our modern world today, we are in crisis faced with many issues that we have not yet found a way to solve – and they keep piling. We have a crisis in healthcare, education, economics, politics, environment, marriage and relationships you name it.

It seems as if every aspect of life on this planet is threatened; things are no longer as they used to be and people are changing for the worst.

How many suicides and murders are you hearing about in the news? There is pollution everywhere; politicians have become the worst enemies of the very people who elect them all in the name of power driven by greed. The world economy is failing, the institution of marriage is struggling to stay, the ordinary man cannot access healthcare and the diseases are becoming a menace by the day.

Something is terribly wrong. It seems that our current world view has reached an impasse. Development may have provided us with various conveniences such as easy access to information (for instance having access to modern technology with search engines), easy travel, effective communication with people from all over the world, advanced healthcare systems and technology, proper clothing and housing (the basic needs) and so on.

One would think that all these things would make the world a better place and give us enough time and peace to focus on our spiritual growth. Unfortunately, these 'conveniences' which have by a great measure, made our lives easier than in the days of our fore fathers have come with a major problem; distraction. We are more distracted than any other generation that has ever been.

In the past, the main distraction was labor; hard labor. People would walk extremely long distances, dig large parcels of land and harvest their seed under the scorching sun and even the 'privileged' philosophers had to write on scrolls with feathers dipped in ink. This was hard work! Today, we type using computers and research any information we need with these super fast and efficient machines. There are machines to till land and harvest and nobody has to walk long distances on foot – they are actually forced to for health reasons.

We have a better and easier life, but what has it made us? We are distracted, confused individuals walking the journey of life out of touch with ourselves and without purpose or direction. There are so many things competing for our attention not to mention the fact that we now live under the illusion that happiness is attached to possessing the nice stuff. As a result, we have become obsessed with getting more and more stuff at all costs. It crazy in this modern world we have and in it is a crazier and lost generation who has lost touch with themselves (their soul).

It is evident that science, technology and the rational mind has failed to solve most of our pressing issues. They have failed in fixing the world and its occupants. It seems that they do not provide answers to most of our questions about how to heal our planet and ourselves, how to coexist peacefully, how to find purpose, how to be happy with who we are and how to live to our fullest potential.

We are now turning to ancient art and spiritual lessons from our forefathers who lived in a 'quieter' era where connection with self and spiritual growth mattered more than buying the next new model of a luxurious car brand. Spiritual alchemy will teach us to keep us in touch with the real self buried under all the distractions and possessions.

Alchemy For Beginners

Let's take the discussion to how we can apply alchemy today.

Chapter 3: Application Of Alchemy

Alchemy is a science concerned with transformation and change. As earlier mentioned, there are two parts to it; the physical and spiritual. The physical is concerned with transforming properties of mater such as metals. The spiritual is concerned with transforming the spirit.

The purpose of spiritual alchemy is to help you invoke and free the spiritual self trapped deep within you, suffocated by the distractions of the modern world and other 'unrefined' parts of self such as fears or limiting beliefs. It frees you from deep hurts, soul loss, fear, defeat, inferiority complex and such like self-destructive personality structures (that you have adapted) so that you can live an illuminated free life true to who you really are on the inside.

The state you gain is best referred to as *'existing as a pure being, free and unobstructed'* – you could also add undistracted. This is the 'gold' of spiritual alchemy.

To get the 'gold', spiritual alchemy attempts to restructure your personality and reform your beliefs. Through it, you can discover your deeper nature as it increases your capacity to discover and relate to your inner-flowing content. You no longer see yourself as defined or taught by the outside world but by what you see and feel on the inside.

It heightens your consciousness and changes the way you view and experience the universe. This is how you get to change your world. If a greater percentage of mankind went through such a transformation, wouldn't we have a changed world – at least for the better? Let's discuss the 7 stages of alchemy.

The Seven Stages Of Spiritual Alchemy

To build something into a new form, you need to 'remove' it from its earlier form. In alchemy, there is a saying for this. It goes, (in Latin) *'solve at coagula'*. Solve means to break down and separate while 'coagula' refers to the process of bringing elements back together (after dissolving) into a new better/higher form. Therefore, the English translation for 'solve at coagula' is dissolve and coagulate.

Dissolve and coagulate is not just a saying; it is a psychological metaphor, which has a deeper meaning; in pursuit of gold, we break down (dissolve) the limiting parts within our being so that we can be transformed into a free wholesome being (coagulation).

In alchemy, breaking down (dissolving) and building up (coagulation) is a process which happens gradually and in stages. You can use the wisdom of alchemy and these magical stages to unearth and defeat limitations within you that have been holding you back and limiting your life. In these stages, you are going to be 'broken down' like the lead and developed into a finer higher quality being, just like the precious gold. There is nothing miraculous about this; there is no foreign power that is going to come and present your best life to you. You are only going to learn how to remove the layers of distractions and delusions to discover that you posses all the power and hear that intuitive guiding voice inside that you may have muzzled – or learnt to ignore.

Let's discuss the stages below.

1. Calcination

In ancient alchemy, this is a process of heating and decomposing raw matter, to remove it from its form, to take a moldable form. Lead was placed in a crucible and heated or exposed to an open flame where it was burnt and reduced to ashes.

Psychologically (or in the practice of spiritual alchemy), the fire of calcinations can be experienced as the metabolic discipline that tunes the body. It begins at the base chakra located at the base of the spine.

Alchemy For Beginners

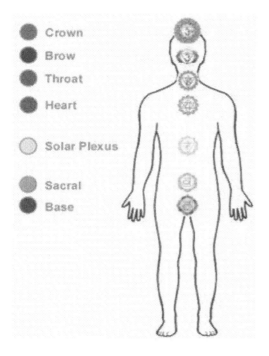

It is through this fire that we break down the parts of our being that stand in the way of our happiness or serving our purpose. Yes, this is the part where you have to accept and look into the real you; the imperfect self, not the 'fine' individual you parade before others. The truth is that most times, in our normal raw form, we live life fulfilling the idea of perfectness that everybody has and neglect what we feel deep inside thereby giving up the chance of being truly content and happy with our lives.

The happiness most of us have is just a mask we wear, but deep inside, we are very sad, scared and confused. Do we ever admit to have such feelings? No. we don't even acknowledge that the feelings are there; the society has taught us that feeling so makes us weak, needy and antisocial, so most of us are out here living a lie, putting on a show that even we have come to believe in.

In our modern world, we get these illusions of happiness, mostly, from likes on social media posts where we present our 'perfect' life in a photo or video. We get carried away and it may take a very long time to realize that we really aren't what we seem to be. Sometimes this happens in old age or in the face of tragedy when our illusion blows away like smoke and we come face to face with our sad realities.

Calcination is a natural humbling process; it strips you off everything you have gotten attached to and every false belief you have held. It involves the destruction of the ego, attachments to and obsessions with material possessions, stubbornness, pride, arrogance and self sabotaging behavior.

For instance, this is where you stop holding up arrogance as a shield from people who make you insecure. Also, it is the place where you strip off the false sense of importance and 'idol status' that your numerous possessions have given you and go back to your authentic self. Yes, this fire eats away anything that is not your natural form and makes you into a natural and freshly alive human being.

It is a difficult stage

Do not expect it to be easy; I cannot promise you that. If it were as easy as it seems, everybody would have done it and be truly happy. Actually, most people avoid it especially when they are exposed just a little and feel the sting; they never go back.

In this stage, you will go through a process that will shock and disillusion you. It starts with deflating your ego. It feels like you are being stripped of your identity and thrown into an abyss of unknowing and your world burned to ashes. It seems as if everything you hold dear is being taken away from you and this may anger or frighten you. This is why most people will feel like drawing back. Your ego is deflated and you can no longer beat your chest about anything; you are back to being a little child who has no influence of the ego.

However.....

It is very beneficial

It is not easy to feel like your world is lost. It is difficult to experience 5 minutes without the ego. However, you must allow yourself to go through this process. It exposes you in all authenticity and rubs away the false self that has been running around making wrong decisions (influenced mostly by the ego) and causing you unhappiness. Yes, it cuts you off from the little charade of living a lie and has you face to face with your naked self.

This can be painful but it is very beneficial. If you really want to make a change in your life, you must change yourself on the inside first. You must reset to begin again; by resetting it, means that you have to start seeing life from a different perspective, changing your thought processes and beliefs. It can only happen if you wipe the slate clean and start over. There is no better way to do this than going through calcination. It gets rid of all that is no longer needed and everything that is limiting you – including some cherished beliefs.

2. Dissolution

When you have burnt matter into ashes in stage one above, you move to the next major step in your chemical process which is dissolution. This is stage where you dissolve the ashes from the previous stage (calcination) in water.

Psychologically, the artificial self is broken down further by immersing self and surrendering to the unconscious, non rational mind. Mostly, this is an unconscious stage in which the non rational mind takes over control from the conscious (rational) mind. The truth is; the conscious mind is shaped by our environment and experiences and it can judge right and wrong and also tell between future and past.

It is the part of the brain that helps you hold limiting beliefs, telling you how impossible it is to achieve something – providing valid reasons based on your circumstances. The conscious has a lot closed within us because of its judgy characteristics.

The unconscious mind however has no ability to do that. When you let it have control, it opens floodgates of strong waters that had been held back. All the material that has been buried is floated to the surface. That's where traumatizing memories that have influenced the development of fear (unknowingly) come to the surface. This is where you face the dreams you gave up on, mistakes you did not acknowledge and therefore did not learn from, a passion you have never paid enough attention to and so on.

It is in this stage that we start to identify less with our false personas and start looking at the real person; the person we often suffocate. Immersing ourselves in the subconscious allows us to become spectators of our own selves. We are able to step back and observe both our negative and positive qualities; we just watch whatever arises from within without judgment, pride or doubt blocking the way.

In dissolution, you are aware of everything you are and everything you have; anger, passion, love or whatever emotion you have been holding. It is here that you see your 'harmless' behavior for what it is; you may have been doing certain things unconscious of how negatively they affect others. I bet this stage can force a rowdy politician to reform when they actually see how ugly their venom looks like.

How does it feel like?

Not good at all. Coming face to face with your ugly realities and the truth about your life that you have been running away from is not going to be feeling good. It is a stage which typically involves feeling lost and crying rivers.

What are the benefits?

We do not give much credit to emotions, maybe because we are oblivious of the fact that it is they that guide our decisions. Think about it, why did you make the decision to marry your spouse? I know you felt love and some other emotions for them, right?

Dissolution brings emotions back to life, while detaching them from experiences that triggered them. This gives us the freedom to feel again. You see, when we connect an emotion to a traumatic event in the past, we tend to suppress it to protect us from re-living a painful experience.

However when we are able to detach emotions from events, we can allow ourselves to feel freely. This is important because the soul is not whole without emotions; they are its main tool of communication. Often, when the soul is communicating, you will not hear a voice, you will 'feel a voice or a feeling'.

We spend so much time and energy trying to hide emotions considered to be dark and unacceptable such as fear or anger. The world is requiring you to smile wide and look happy and to be fearless while on the inside, you are feeling different. If you allow yourself to reintegrate the emotions, you would be surprised at how whole you would feel. The truth is; most are the times, we feel like something is wrong with us for experiencing certain emotions – so we refuse to feel them. However, it is important to note that emotions do not go away without being acknowledged.

Rejected and hidden emotions have the potential to grow dark and dangerous with bad consequences. For instance, if you hide anger, it can become loath and before you know it, a person becomes suicidal or is bent on seeking serious revenge for being hurt – and we are talking homicide kinds of revenge which are very common nowadays.

We need emotions whether good or bad to feel alive. Living in fear of experiencing an emotion is simply living afraid of being alive, trying to be perfect, feigning happiness. Dissolution will bring them to the surface, the good and the ugly (pun intended) to help make you feel whole gain.

To say the least, dissolution makes you aware of your avoidance patterns, brings whatever you avoid to the surface so that you can really see it and deal with it. It just gives you a much needed wake up call.

3. Separation

This is the third 'operation of transformation' in alchemy. As a chemical process, it is done by isolating the components of dissolution (the ones dissolved in water) by filtering the solution. It is meant to separate the good and genuine materials from the waste (unworthy material).

Psychologically, this is the stage where we learn to make better decisions, by separating emotions, thoughts and actions. Our true feelings and emotions are more defined and we can separate them from the false ones that are imposed on us by peers, environment or influenced by history.

Separation helps us become aware of our authentic feelings towards a person, situations or for ourselves. For instance, if you are angry about something, you honestly experience your anger instead of shoving it aside to be happy about something because it is the right or socially acceptable thing to do.

How does this help?

Most of us are confused about our character feelings and thoughts. We are entertaining people who do not add any value to us or who are hurting us because we are dutifully forgiving and forgetting and accepting things which are not good for us maybe because it is more comfortable to do so.

This is happening because we are not allowing both sides of our emotions to come to the surface. Instead, we weigh more on what we think or what the society considers to be right. For instance, when we love someone who hurts us, we need to bring both emotions; the love and anger, to the surface side by side so that we can make informed decisions based on which one weighs heavier. You see, people who stay in abusive relationships will often look at the love emotion because everyone expects them to stay put because they chose that person in the first place and we all hate to have made a bad choice, don't we.

'But I love him/her' is the classic mantra and choose to ignore the other emotions arising from their pain such as anger, resentment, and hatred which perhaps would make them see the reality of their ugly situation.

All emotions count. If you try to suppress any one of them, then you are going to find yourself being indecisive and confused about situations that would otherwise be so easy to maneuver. Our souls know best and they give us the right direction. The soul and emotions are interconnected; the soul expresses through emotions. Recognizing and understanding your authentic emotions is connecting with the soul and anyone who lives guided by the soul is living their life to the fullest, being and doing what they were created to be and to do respectively.

Bottom line is, separation helps update our old ideas, beliefs and thoughts. Once dissolution gives us a wakeup call, separation helps us align our emotions with expressions so we can live in the truth of now. For instance, as we saw in our example about abusive relationships, separation will update your emotion of love (which may have been there when things were rosy) with the current experience where you are feeling anger and hatred now that things have turned for the worst and you will be guided to make better decisions for self.

4. Conjunction

Alchemists believe that the things of the earth are greatly influenced by the things in heaven, so they came up with a saying, "As above, so below". They were bent on seeking balance between the two so as to effect healing and transformation.

They believed that to be whole, we must have a connection with heavenly things and at the same time be rooted in the earth. The idea is to have a highway within where the two freely flow to and fro.

What's the significance of this?

The heavenly things represent the unconscious; the things that seem to exist in another dreamy and higher place. The earth things stand for the heart and the earth. Conjunction is a stage whereby the spiritual (from the unconscious) rises to the surface to meet the physical (conscious).

Just the word is used in English grammar for words that connect others to make a sentence; it is a stage where everything that was achieved in the first three stages (which involve purification and clarification) is connected.

In this stage, there are lesser fears, confusion and baggage. Our unconscious thoughts and feelings have come to the surface and are integrated into the light of conscious awareness. Now, we can connect the dots of who we really are and move to acceptance of our true authentic self. It is a time to create a new more live being, authentic and living in purpose. So, we create new attitudes towards self, other people and situations.

In separation stage, you identified what you are supposed to pursue and what is no longer worth your effort. Now it's time to place yourself on the path you feel you ought to be on; it's time for you to engage in what you have always wanted to do. Start doing the research, socialize with like - minded individuals, and look for the appropriate social connections and so on.

Remember the 'should not's' have cleared off and there is neither self-sabotage or limits. It's time to set the stage to fly as high as you can dream of.

Experiencing conjunction or in other words, a 'conjuncted consciousness' brings us closer to being the person we believed we could be; the person we came into this planet to become and serve the purpose which was set aside for us. However, this is not where it ends; you are not 'there' yet. Look out for the next step.

5. Putrefaction-fermentation stage

a. Putrefaction

The ego will attempt to dominate and start with its little lies of why this or that cannot be done, giving you valid reasons why you cannot do certain things meant to elevate you. This is stage meant to deflate the ego and disable its sabotaging tendencies. Therefore, you go through Calcination- like process only this time the experience is more intense.

Also, you are not alone scared and lost like you were in the first stage. Now you have the support of a higher force within, a spiritual source of strength from the soul.

Putrefaction will not be a smooth sail however. It can be very disturbing, especially after you thought you had fixed everything. You may feel like you are sinking into an abyss of despair and depression. However, you have the strength illuminating from your soul to help you get through this and you feel supported, even encouraged, watching objectively and trusting the process. This stage can be perceived as the death of a grape to enable the birth of wine or the death of a caterpillar to birth a butterfly.

The ego is straightened out in this process so that it does not go rogue or try to sabotage us. It is made so that it supports and gives us the strength to push forward. It is only when we have been refined enough through the intense fire of this process that we can go to the second step of this stage.

b. Fermentation

This is the stage that marks our rebirth where we can start enjoying the fruits of our new refined self. The other stages involved working with aspects of our old self but here, none of those exist; we are working with the new empowered individual.

In ancient alchemy, the components harvested from all other stages are transferred into a new container – to mark a new beginning. A catalyst is added to help push the final stages of fermentation where a new substance is formed.

In our reality, our egoless surrender to the process symbolizes the empty vessel. The catalyst may take different forms but it must include new insights, perceptions and an enlightened view of self, others and our environment to foster the creation of a new life/reality.

In this stage, we are aware of why we act as we do or why we feel a certain way, which means we are in control and nothing about us catches us by surprise anymore. Therefore, we have complete control of our actions and we are capable of influencing our outcomes.

6. Distillation

In chemical processes, this is a stage where the fermented solution is boiled and condensed with the aim of removing any impurities which would have been left behind. Psychologically, distillation occurs for the same reason – to give the soul and persona the ultimate cleanse.

In this stage, you go through a process which ensures that there are no impurities carried forward from the previous stages or your past life. For instance, it further checks whether the ego has fully converted from being inflated and sabotaging to being all supportive. Basically, it is a level of further purification to make us into the enlightened being we all should be.

An example of the distillation process in our day to day life is practicing to live as the five stages above have taught you to. For instance, you practice to live and perceive life from a daily place of peace and positivity even in the most trying times when being otherwise would be excusable. You could be wronged by someone very close to you, and they know that they did wrong and would not judge you for being mad and making a fuss.

Instead of what doing what the former you and everybody else expects to do, you respond with calmness, acknowledge that you are angry but you let the emotion flow through you without attachments – for it is in identifying with an emotion that we give it power to influence your actions.

Distillation is a stage defined by a whole lot of practice and 'forcing' yourself to become the individual we have worked on in the earlier stages. It is not going to be easy to change, especially after sticking with some habits for a very long time. You have to do enough practice of being the new being, as if you died and rose again into the present without reverting back to the former habits and identifications - those are no longer yours.

If you find yourself slipping back, there is nothing for you to worry about; you just need to keep practicing hard. Eventually, you will experience a strong, profound transformation, what the Buddha would have referred to as Enlightenment. This stage is meant to address those shortcomings and refine you to become the finest.

A completely distilled person is different from anyone who has not been here. There have a rather mature perception of things. For instance, they do not live life with some fairy tale expectations influence by the ego. Also, they do not think that life is miserable (and go to extends of taking their own life as are most cases today) just because their expectations were not met. They work with and are led by the soul, and their ego is supportive of this. They surrender to higher forces (God) to lead their lives while maintaining respect for 'smaller forces' (discipline, hard work, family and so on) that shape their life.

7. Coagulation

This is the last stage of alchemy; it is the stage where wholeness is confirmed. It is not guaranteed that you will move through steps one to six and get here with a onetime try. Most people who have reached coagulation have gone through the stages, especially the lower ones of burning obstacles (such as the ego), redeeming emotions and making a decision about what matters many times. Most people do not get past these lower stages because it is hard to move from the lies that make up our 'perfect' lives and your little perfect world torn apart such that you feel lost with no grounding.

You are not special either, and may have to go through them many times before you move on to discovering your heart and yourself, living the life you have always wanted and practicing to master how to live like that every single day and finally getting the ego to be on your side. It is until you have mastered these in all the former 6 stages that you get to coagulation where you are confirmed to be whole.

So can you rest now? No. coagulation is not the end. They say that getting to the top is easy, what's hard is maintaining your position there. Same case applies here. You have to fight and practice every day; now you live a day at a time and live it right. You cannot rest; you require more responsiveness, intention and careful tread, listening only to the soul so that the ego never creeps back to its old controlling and sabotaging habit again. There is more discomfort, at least until you adapt to your new way of living because let's face it; old habits are a comfortable place we were used to and easy to go to. You also have to be mindful than ever before on your moment to moment decision as now you are fully aware that they are going to affect the universe; your universe.

Few people ever get to the point of full coagulation – almost none actually. We are all a work in progress when it comes to finding the gold within us. Out here, there are so many dark forces and negative energy that is bent on tainting us, with the ego being our biggest enemy. We may fall victim from time to time but this does not mean that we stop; we do not stop when we fail, we fight to get back to the top, being wholesome beings. It is inevitable that we must continue to go through the process (all the stages) over and over again to continue with the refining. Even though there is a lot of toiling, failing, pain and discomfort, toil we must like the miners do to find the precious metal. The struggle is not easy but when you see the quality of your life rise, it will be totally worth it.

Chapter 4: Effecting Transformation With The 7 Stages Of Alchemy In Modern Life

We only discussed the stages and what to expect in the previous chapter. This is the part we discuss how to actually do it. In this chapter, we shall be discussing doable techniques, which we can use to perform the processes and achieve the desired results of each stage. In other words, we are making modern alchemy actionable and learning how to incorporate it in our daily life and routines.

Let's get started, from the first stage:

Breaking Down The Present Self

1. Undergoing Calcination

This is the first part where you go through breakdown of self i.e.bringing down the little world you have built around yourself. It involves bringing apart your comfort zone and interrupting your 'usual' thought processes. Of course, I will not ask you to go get heated in a crucible or over a flame, but somehow, you have to go through some form of heat, like really looking and paying attention to what's going on in your mind or your life. You agree with me that we do not do that a lot – we do not have time to.

Maybe it's because it feels uncomfortable, especially if you have to look and come face to face with the lies, truths that illicit uncomfortable emotions.

So, how can you put yourself through this heat?

Actionable steps

- *Reflect on your life*

Pick a day when you do not have so much to do, when you are feeling a lot relaxed and can have some time by yourself. It could even be in the evening, after your busy day when you have tucked yourself in bed, warm and relaxed and nobody is going to burst in your room and there are no distracting noises from your environment.

Close your eyes and look back at your day or the past few months – just as much as you can handle but little bits are recommended because when too many 'realities' hit home at once, you may just lose it.

Remember to look at situations for what they really are/where – nobody is judging and it's a moment safe from social media.

Examples of questions to ask yourself

✓ Did I really do the best I could today?

✓ Did I give my best self to every person (even the most annoying) and situation I went through?

✓ How real has been my life/relationships? Do I really like the way I am living or being in this relationship or what motives do I have for being here?

- ✓ What is my real life away from friends, social media, spouse/partner, and work like?
- ✓ Am I truly happy?
- ✓ Do I regret anything I did and why?
- ✓ What do I want? Is what I am/have been doing leading me to what or where I want to be?
- ✓ Am I living the life I have always dreamed of? If not what has been limiting me?
- ✓ Do I like myself right now? Am I proud of whom I have become?

You know what your life is like and you know where you need reality checks. There could more questions to ask. The aim is to reflect on your life, what it has been and what it is- the behind the scenes raw story that only you know.

Why reflect?

If you want to make a change in any area of your life, you have got to have enough reasons why. When you reflect and see the real story of your life, if at all how you are living or your circumstances aren't anything you are proud, you will consciously recognize that you need a change. This marks the beginning of transformation; the recognition that things are not okay and you need to be or do something differently.

Is it all going to 'come to light' instantly?

You are not going to sit for 20 minutes reflecting on on single occasion and figure out everything. You may not even be able to fully focus – and that's okay. What's important is that you commit to the process and make it a habit to reflect every day. 20 minutes every evening or morning (whenever you can be relaxed and able to focus) will certainly cut it. Do not rush the process either, give yourself time and put in a lot of practice. The insights will come eventually; after going through the heat and fully evaluating your life.

It's going to hurt

It does not feel good to evaluate the raw details of your life. We live in a fast world in an extremely fast way and sometimes so many lies, half-truths, pretences, false personalities, false successes and many other ugly things find their way into our lives. In fact, if there is one thing common among human beings is the disease called delusions of grandeur. We think ourselves to be more important and make a great impact than we really are.

Most of us are suffering from this disease. For instance, look at how we post 'success' on social media while in real life we are struggling with our careers, our lives and are unable to make ends meet. However, for some reason, we have convinced ourselves that we are 'fine' and doing well thus we can afford to go hang out at expensive places just to please friends and have some cool photos to share. If not this, we 'e time on the internet reading stories sful superstars) catching up on what's of focusing on what we need to do to We forget that those superstars are really is is as a result of putting in work, us and we are just reading and starring hours on end at pictures of their success. We, on the other hand, haven't built anything but being up to date with everything that's going on in the world makes us feel important, even though we may be living from hand to mouth. This is sad but unfortunately, it is the reality for most of us today but we just never see it that way.

Without pausing to look at your life, you may never recognize such things that are inhibiting your growth. When you finally see how pathetically you have been living, it will devastate you. Coming to terms with your truth that you are nothing as you have lied to yourself you are for many years is going to make you feel lost and could even affect your self-esteem.

It does not feel good, at all. However, it is good to feel this lost and this sad because it will bring on the 'ah-ah' moment when you know that you have to do something differently. It is the beginning of your transformation to become better. Yes, it hurts and burns. There will be lots of tears. Do not let that stop you from diving even deeper to look at your life as a spectator would, seeing everything for what it is. The fire of reflection is hot but necessary – and totally worth it.

Methods of self reflection

Self-reflection is not limited to one method; there are several methods to do this. Use whichever works best for you.

They include:

- ✓ Meditation

- ✓ Deep prayer; speaking to a higher spiritual authority

- ✓ Talk to yourself about a subject you would love to reflect on. In other words, think out loud. Things make more sense when you hear yourself speak about them.

2. Going through dissolution

You have come face to face with the good and also the painful realities of your life, so what next?

Immerse yourself in the unconscious mind. You have come this far depending on the guidance of the conscious and rational part of the brain, which in many instances, is affected by the environment. This part of the brain has limitations. Even worse, it is under the control of the ego. The ego is fond of instant gratification; it likes to be pleased and would rather achieve pleasure than do what is right or beneficial in the long run. This said, you already know that it will sabotage your resolve to achieve a long term goal and have you doing unimportant things that makes it happy now.

The unconscious mind has no limits and is not controlled by the environment or ego. Also, it is the best at storing information about you; it knows everything you have been through and every emotion you have ever felt so it can connect experiences with emotions. This is where you can find and reconnect with buried memories and thoughts that may have affected your life in one way or another.

The answers to who we are right now and what our journey here has been cannot be found in the present. The past contributes greatly into making us. Childhood experiences can shape your adulthood. For instance, if you were overly criticized as a child, you may find that you have self esteem issues, fear or lack of belief in yourself which can be manifested in your inability to speak your mind (and you get used a lot without ever getting what you want) and lack of confidence, even in brilliant ideas you have. You could be wondering why you never seem to make any progress in your life or career. You have read every self-improvement book available and even seen numerous counselors but nothing seems to fix you.

In such cases, you need to go through distillation; connection with your unconscious mind. In there, you will find answers. You can connect the dots backwards and find links between experiences and emotions so you can be able to understand who you are and why you are the way that you are today.

Actionable steps

Observing your mind meditation

- ✓ First determine what you want to accomplish. Do you want to retrieve memories of your childhood or reconnect with a lost dream? Determine what you want to gain in the process before you get started and focus on that; otherwise, you will be swamped by all the information in the large storage called the unconscious mind.

- ✓ Forget about your consciousness; stop trying to control your mind. Just get into the flow and ride with the tides of your brain. You can do this if you completely let go of all control and just let yourself float in nothingness.

- ✓ Forget about what your body needs – trust me, this is the time when funny needs like itching, thirst and so on are going to pop up. Do not try to snob the wants; acknowledge them and in your mind's eye, watch them pass as you return your attention to your mind. Watch your thoughts arise without trying to dwell or judge any of them; just keep watching as a spectator and see what arises from the unconscious, noticing every thought, emotion or feeling.

3. Separation

After successfully going through step one and two, at least now you have an idea of who you are and what you need. This is the information you use in the separation process where you make conscious choices of what you need to keep (whatever supports the person you are and the goals and aspirations you have) and that which you need to discard (whatever does not add value or support your present and future self). Through separation, you rediscover your authentic self and create a favorable environment within to become who you want to be.

Actionable steps

- ✓ Have some self-discovery sessions with a journal in hand. In these sessions, you get to think about what you want and things you feel that you need in your life. Here you are not thinking about your best friend, your spouse or kids. This is about you and what you want. Do not think of it as being a selfish move as such thoughts can stand in the way of a successful session. Think of it as moments taken to understand and fix yourself so that you can become a better spouse, parent and friend – they say you cannot pour from and empty cup. This is a process meant to fill your cup with love, passions, understanding you name so you can be able to share with others.

How to do this

So, this is about you. Think about you; what do you want, what do feel that you need in your life at this point and time and write it down. Writing does not only help you keep a record of your thoughts but it also a great way to gain insight and become self-aware. You will see that as you write a single thought down and try to express yourself in the journal; it feels like you are connecting with yourself at a deeper level.

Why write it down?

A journals main job is to give you a reference point where you can check and make a connection between the way you think and the way you behave. You see, sometimes we have very good thoughts; brilliant ideas, goals and a smart course of action planned in the mind.

However, when it comes to execution, we do thing very differently and end up with dump action plans, changed goals and we forget how brilliant our ideas are and adapt those of others (or whatever we are told is more popular). This is the reason most of us are great thinkers but miserable failures.

When you write it down and constantly refer to it, you are kept in the path of what you ought to be pursuing. In moments of doubt or when you face tempting distraction to stray from your goals and values, you can always revise and give yourself a wake-up call or get back in line.

✓ *Determine your values*

What are your values? They refer to the things/standards/morals that you consider important in life; things that you have a high regard for. To determine what yours are, you can try thinking about characteristic that appeal to you most in other people – people that you admire and respect. Why do you think those characteristics stand out for you? Maybe, it is because you consider them to be important and if you do not already have them, you would want them in your life. They could include something like self respect, discipline, good morals and so on.

Also, think about issues or events that you consider to be important to you. It could be a nature walk you once had with a friend and you felt so relaxed and peaceful. This could direct you to suitable places for unwinding (instead of trying to relax with TV shows), the type of friendship you value or a person you should have in your life (if you somehow disconnected).

✓ *Recognize your shortcomings*

We all have good qualities and we should focus on growing these ones the most. This is not to say that we should forget that we have negative ones too – which we can often be very blind to.

Identifying your bad qualities and shortcomings is paramount to the success of your separation stage. You see, the good qualities will not stand in the way of your growth to a higher being, but the bad qualities are the enemy; you need to know your enemy so you can devise ways to clear them from your path.

It is not always easy to recognize your shortcomings. Usually, we are perfect in our eyes but we are very aware of other people's faults. Therefore, you can use others (who are very aware of your negative qualities), more so people who are close to you who would offer positive criticism, to point out those shortcomings. Maybe you are terrible at keeping time and they always have to suffer for it – but buried in your believable excuses, you do not see it as a problem. It could be that you do not keep your word and this could be the reason why you do not achieve your goals.

Find a few trusted friends or family within your circle to give an honest assessment – of course after you have already done yours. This will give you an insight into what aspects of your personality you should get rid of and which ones you could improve on. This move will help shape you into the higher being we are helping you become.

Creating A New Self

4. Getting conjuncted

Here, you are going to assemble the saved elements of self from stage one to make you into a new being. As mentioned earlier, you are going to bring together the conscious and unconscious mind and the feminine and masculine elements of your personality to create something new in yourself.

This 'new thing' could be a new vision of self, a new personality, new beliefs, a fresh view of the world or a new state of consciousness and so on. The important thing is that you bring together elements within you to create this new; do not try to 'ship' elements and attempt to become someone you are not may be because you desire to be that way. This would be an attempt to create another false self after you got rid of the old one – it would only bring more suffering.

So, it is best to become the best version of who you already are – you are already enough.

Actionable steps

- ✓ Meditate to access your unconscious mind and bring it to the conscious level

- ✓ Visualize; close your eyes and see yourself in your mind's eye being the best version of yourself. Imagine that you have got everything under control, doing what you are supposed to and achieving your desired goals. See yourself being happy and emotionally in control. Think of the things that matter most to you and see them draw to you. Think about your positive aspects and like a bright light, see yourself use them to illuminate your world.

Visualizing helps bring the unconscious desires to the conscious world so that they can be actualized. There is a wise saying quoted from the holy book, which goes, 'Visualize things that are not as if they were…' for them to manifest.

5. Fermenting the new self

When you have won against the ego and gotten over limiting aspects of your life and have transformed to this new being, you will be happy and loving the journey. We all hope that we have settled into a smooth sail being at our new level of consciousness, happiness and success.

The truth is you still have to keep on with the fight to stay how and where you are right now. You have to take your new aspects through a process of fermentation to make them stronger and sustainable. Here, you are no longer trying to find yourself; you already know who you are and what you want. You are making your new habits stick, moving away from the possible pull back to what you formerly was.

Actionable steps

You need a means to connect you to a higher focus and higher power to help and sustain your transformation.

You can achieve this through;

- ✓ Deep and intense prayer – this is meant to give you strength and reassurance in your purpose in moments when you feel weak or when you are facing disillusion.

- ✓ Getting spiritually connected to a higher power – it assures you that you have a power greater than yourself to hold you up and be accountable to.

- ✓ Deep meditation (done frequently) – it gives you access to the unconscious mind which you need to constantly draw from in your new state of consciousness.

- ✓ Transformative therapy – it helps with transitioning you to your new state.

6. Getting distilled

This is a purification process, to remove any impurities that may have seeped in. For instance, from the wine that was made into grapes (as mentioned in fermentation), when distilled further, we can make something more pure like brandy by boiling off an impurity, which in this case is water.

In your transformation, distillation is where you expose your personal psychic forces to a higher level of refinement to ensure that no elements of your former self have remained.

Actionable steps

- ✓ Look at your life from the view point of a spectator

We can be quite self righteous and think the best of ourselves and our actions when we look from our own viewpoint. We have an inner explanation and excuse for everything otherwise termed as we understand ourselves. This is not the case for others; they do not understand us. Instead, they see us for what we are and whatever we do and make conclusions about our personality.

This is a little bit harsh, right?

This is the reason why you need to look at you, as if you did not understand yourself – like the spectator. Question your character, your actions, beliefs and thoughts. What do you think about them? Are they reflecting the person you are trying to be? If not, recognize the chaff, get rid of it and use that critical eye of the spectator to look again until all the impurities are removed. Some of those impurities could include (these ones tend to hold on too tight for far too long); a deep seated fear of the future, self-doubt, ill motives, misplaced priorities, bad habits such as procrastination, limiting beliefs and crowd pleasing.

✓ Perform an objective personal introspection

Introspection is when you 'think about it'. You think about your thoughts, emotions, behavior, motives and actions to develop a higher awareness of what you do, how you tick and also get an insight on your perception of the world.

During this process, you ought to ask yourself questions such as;

- Why did I react that way?
- Why am I feeling this way?
- Why am I doing this?
- Why is this bothering me so much?
- Was my reaction called for?
- Why do I need to do this?
- Why am I thinking about this? And so on

These kinds of questions help you sort things out with and for yourself. You can do it either by yourself in your mind, while journaling about your day (highly recommended) or in a talk therapy session with a counselor.

It is meant to keep you in touch with yourself and how you project aspects of yourself so that you are aware of what is really going on with you and if there is something you need to change.

7. Coagulation, finally!!

This is the part where you are finalizing the change within. The work on you is finished and now you can be released from our 'alchemy lab' into the world as a new completely transformed being. This is a stage marked by a profound confidence in self and an ability to continue to exist as the new being with the amazing qualities you just found within yourself, in any aspect of reality – without being turned.

How does it feel to be here?

You should be experiencing a sense of higher being and feeling rejuvenated. You have been through a lot as you went through the process; there probably was a lot of discomfort, pain tears. However, at this stage, you ought to feel a new freshness, like those tears were scrubbing off the dullness and unhappiness that you had experienced before the Enlightment.

The hallmark of this stage is confidence, commitment to live purposefully and emotional balance all of which represent a higher sense of being.

Chapter 5: Keep Your Alchemist Practice Alive And Maintain The New Sense Of Self

Getting to the last stage and becoming transformed should only be the beginning and not the end of your alchemist practice. Make it your way of life, a daily practice so that you never lose the person you have been transformed to and the fresh scent of life that you have began to sense.

We live in a crazy over-filled modern world. There is a lot of external noise, with things that are trying to steal our conscious attention. It is easy to get lost in the maze and lose yourself.

How then can you sustain your alchemical practice or maintain your enlightened state? You do only by focusing inward when you turn your attention inward and work on improving yourself; there is little that the noisy external world can do to interfere or distract you.

The following strategies will help you with this, clearing your aura (energy field) and developing a safe circle in which it can exist with positivity, protecting it from external harm.

- *Focus on your inner world: Begin and end each day in your inner laboratory*

Let's describe a typical day in the life of an ordinary man; it starts with reading news, texts and emails, or catching up on trends and the fresh gossip on internet media and ends in a similar fashion. If not that, they will miss the alarm, wake up late, jumping straight out of bed, into the bathroom and out of the house in minutes and finding themselves on the work desk dealing with the days stresses not to mention office drama. Next thing is being stuck in traffic in the evening rush hour, getting home tired, having dinner and sometimes too tired to bathe then jump in bed, scroll the internet and doze off and the gadget drops right beside him where it will be picked in the morning to browse as a morning ritual.

How then will a person living like this not be lost in the external world if they constantly feed on information about it? Their life is guided by the outer world and all the trash that is being dumped in it.

As enlightened as you are, living like the ordinary man is going to cost you this new being that you went through fire and pain to create. It is important to remember that you are not immune to the noise and influences of the outside world. In order for it to have a minimal effect on you, this is how you should start and end your days;

Start the day by looking within;

- ✓ Meditate to connect with your spiritual being and the unconscious every single day.

- ✓ Visualize how you would like your day to turn out – and make sure to keep it positive because what you visualize and put in your mind manifests. Say for instance, you are going to negotiate with a mean client. Visualize that your meeting will go well and that you will manage to strike a deal.

- ✓ Connect with a higher spiritual being. For example, you can pray and seek guidance and pray that your day be illuminated.

End by looking within

In the evening, just before you retire to bed;

- ✓ Journal how your day was and record the events that stood out for you and also your wins nad failures alike; this helps you celebrate you while still noticing where you went wrong or where you ought to make a change.

- ✓ Write down what you are grateful for (you should have a gratitude journal)

- ✓ Read a chapter of a good book

- ✓ Carry out some personal introspection to check in with yourself; examine your thoughts, actions and reactions and feelings.

- ✓ Pray and be thankful for the day and let your mind, body nad soul relax

- • ***Isolation***

Ancient alchemists did their noble work in secrecy; they kept it low key. Also, they focused more on spiritual connections than social connections, as they tried to make a breakthrough in their work.

Given that some part of the population did not believe in alchemy, they did not hang around anyone who would trash their work. If we could follow with their social interactions, it is likely that they kept the company of other alchemist, scientists and alchemy enthusiasts.

When it comes to your social life, it is wise to emulate the ancient alchemist. Therefore:

✓ *Keep a healthy circle of friends and associates*

They say that you are an average of the five people you spend most of your time with. Who are you spending time with now that you are a new being? Is your internet gossip addicted friend still coming to sleep over in your house? Do you still go for weekend getaways with that negative friend?

Make a decision to spend time with people who respect you, support your journey and more so who have goals similar to yours. These are people who are going to support and hold you down in weak moments and remind you who you are. When you have similar goals, at least you will be walking the same path, sharing knowledge while giving each other support.

✓ *Develop social boundaries*

You do not have to hang out with people or be in social places that you would rather not be in – people and places you feel are draining your energy. Go only to events that you want to and comfortable in. Engage in activities that you feel are adding value.

Hangout with people who give you positive vibes while limiting as much as you can interactions with people who tempt or threaten to lower your vibrations with their negative behavior. This is because your environment and the activities you put your heart into have impact on your vibrations/energy. At this point, you should be trying to keep your energy as clean and positive as possible.

✓ *Censor/limit your exposure to media*

Media content has been specially created to grasp our attention and influence our thinking and behavior. This is why you should be careful with the type of media you are exposing yourself to and also how much time you spend on it.

In our world today, everyone is trying to make money on the media and thus you cannot be assured of quality content, because all sorts of stuff are being put out there. If you want to maintain your transformation, you should limit junk viewing; lest you 'accidentally' feed your brain content that corrupts your new thinking and inhibits your mental and spiritual growth.

It is important to predetermine how much time you are going to spend feeding on media content say on television. Select the shows you are going to see; especially the ones that are informative or those that bring you joy. Get to know at what time they are aired, watch them and get away from the TV when they are done. However, you cannot spend 5 hours trying to relax with fun TV shows – this is binging on entertainment and it might end up draining you instead.

- ✓ *Limit your time on the internet and feed on constructive content*

It's obvious you have internet access and your smart gadget everywhere you go and logging on to social media sites (the epidemic of our generation) is easy. It can be very tempting; we get entertained on the internet, we show off and get validation (with likes) and get distracted for hours on that world were everything seems perfect when our realities get tough.

You realize that unlimited entertainment and all those distractions will drain your energy as you get drawn into the fantasy world thereby losing consciousness of the real world; your reality. You lose touch with reality and in the long run compromise your mind, thoughts and emotions as they get influenced by an unreal world. In addition to wasting a lot of time, you set your mind and consciousness on paths that do not serve you or your spiritual growth.

So, use the internet for causes that benefit you. It is okay to get some entertainment there but remember you can also get access to a lot of beneficial information online. How about you entertain and unwind for one hour or so and commit the rest of the time educating yourself about important stuff;

there are plenty of educative blogs and v-logs online about business, spiritual growth, healthy eating and living and so on.

- *Prioritize alone time*

You need time alone; it is very important to set aside time to have quite time alone. It is during these moments that we can wear our emotions on our sleeves, give ourselves assessments for our spiritual and personal growth and check in to see how far we have come and how much we need to improve on. You cannot do this if you always have people around you all the time.

There is no way you can let your deepest emotions and feelings come to the surface say in front of your friends or family. You know that some of those are not pretty and sometimes you may need to get too emotional and perhaps shed a few tears. If you do not give yourself time and space to do some of this stuff, you will find that you are bottling up too much of what you ought to let out. You will be making your inner laboratory toxic – and you know how big a role it plays in keeping you in the new found heights of consciousness.

- **Get rid of/avoid clutter**

Too much stuff not only draws on our attention but it also keeps us emotionally tied to possessions. Believe it or not; every component of matter has an energy field around it. Whatever we keep around us we link their energy fields with ours through invisible cords. So, what happens if there is so much linked to you, especially of stuff that you do not need or feel a connection with? You will get overwhelmed and constantly feel spiritually and emotionally drained.

This is not to say that you should not own stuff. You can have stuff but only that which has significance in your space; something that you need or love. Also, keep it organized; disorganized stuff has a way of overwhelming vision and the brain too. What's more; it's important to let go our attachments to our possessions because this gives them power to own and control us emotionally.

You would be surprised at how clutter free spaces leads to mental peace. They give our brains less stuff to work through by simplifying what we see and also by reducing the amount of things we have to go through before we can get a specific item we need. This is very important in keeping our brains functioning at their highest level for mental wellness and productivity. Have you wondered why the most successful, highly productive people tend to wear the same kind of clothing? They know that clutter can cause them brain fogginess and emotional confusion – they make a lot of important decisions and they cannot have such. That's why they keep it simple and clutter free with their stuff. It is a good habit to copy.

Conclusion

We have come to the end of the book. Thank you for reading and congratulations for reading until the end.

Ancient alchemists may have failed to grow basic metals into their highest form. However, their research was not in vain. They did leave a legacy of principles of growth which teach us to not write ourselves off just yet because of how 'basic' we may seem. There is greatness within us waiting to be unveiled, only that it takes a lot of pounding, heating and cleansing. Regardless of the pain and strain, we should always trust the process which will eventually convert us to the best versions of ourselves; gold. It was tough, they were criticized, they failed but they never gave, so we should not.

If you found the book valuable, can you recommend it to others? One way to do that is to post a review on Amazon.

Do You Like My Book & Approach To Publishing?

If you like my writing and style and would love the ease of learning literally everything you can get your hands on from Fantonpublishers.com, I'd really need you to do me either of the following favors.

1: First, I'd Love It If You Leave a Review of This Book on Amazon.

2: Grab Some Freebies On Your Way Out; Giving Is Receiving, Right?

I gave you a complimentary book at the start of the book. If you are still interested, grab it here.

5 Pillar Life Transformation Checklist: http://bit.ly/2fantonfreebie

PSS: Let Me Also Help You Save Some Money!

If you are a heavy reader, have you considered subscribing to Kindle Unlimited? You can read this and millions of other books for just $9.99 a month)! You can check it out by searching for Kindle Unlimited on Amazon!

Printed in Great Britain
by Amazon